U.S. Department
of Transportation

**National Highway
Traffic Safety
Administration**

www.nhtsa.gov

DOT HS 811 712

February 2013

Effectiveness of LED Stop Lamps for Reducing Rear-End Crashes: Analyses of State Crash Data

DISCLAIMER

This publication is distributed by the U.S. Department of Transportation, National Highway Traffic Safety Administration, in the interest of information exchange. The opinions, findings, and conclusions expressed in this publication are those of the authors and not necessarily those of the Department of Transportation or the National Highway Traffic Safety Administration. The United States Government assumes no liability for its contents or use thereof. If trade names, manufacturers' names, or specific products are mentioned, it is because they are considered essential to the object of the publication and should not be construed as an endorsement. The United States Government does not endorse products or manufacturers.

Suggested APA Format Reference:

Greenwell, N. K. (2013, February). Effectiveness of LED Stop Lamps for Reducing Rear-End
    Crashes: Analyses of State Crash Data. (Report No. DOT HS 811 712). Washington,
    DC: National Highway Traffic Safety Administration.

1. Report No. DOT HS 811 712	2. Government Accession No.	3. Recipient's Catalog No.
4. Title and Subtitle Effectiveness of LED Stop Lamps for Reducing Rear-End Crashes: Analyses of State Crash Data		5. Report Date February 2013
		6. Performing Organization Code
7. Author(s) Nathan K. Greenwell		8. Performing Organization Report No.
9. Performing Organization Name and Address Evaluation Division; National Center for Statistics and Analysis National Highway Traffic Safety Administration Washington, DC 20590		10. Work Unit No. (TRAIS)
		11. Contract or Grant No.
12. Sponsoring Agency Name and Address National Highway Traffic Safety Administration 1200 New Jersey Avenue SE. Washington, DC 20590		13. Type of Report and Period Covered NHTSA Technical Report
		14. Sponsoring Agency Code
15. Supplementary Notes		

16. Abstract

The purpose of this report is to analyze the crash-reduction benefits of LED stop lamps and LED center high-mounted stop lamps (CHMSL) using real-world crash data. Previous work on this subject included laboratory experiments that suggest LED lamps were more beneficial than incandescent lamps at preventing rear-impact collisions. NHTSA statistically compared the overall ratio of rear-impact crashes to a control group of frontal impacts before and after the switch to LED. Overall, the analysis does not support a firm conclusion about whether LED stop lamps and LED CHMSL are more effective than incandescent lamps. The main analysis shows a significant overall 3.6% reduction in rear-impact crashes with LED. On the other hand, a non-parametric analysis not only fails to show improvement in significantly more than half the models, but actually shows an increase in rear impacts with LED for 9 of the 17 make-models that switched to LED. It was just the favorable results for high-sales vehicles such as Honda Accord that pulled the overall result into the plus. Furthermore, and perhaps most important, none of these 17 make-models is a "clean" switch pair that shifted to LED without changing anything else. All of the switch pairs shifted to LED at the same time that they changed the rear-lighting configuration and/or redesigned the vehicle. Basically, the crash data probably won't support a firm conclusion until we have more switch pairs, including some "clean" switch pairs.

| 17. Key Words
NHTSA; NCSA; State Data System; effectiveness; evaluation; statistical analysis; automotive lighting; center high mounted stop lamp; CHMSL; rear impact; crash avoidance; stop lamps | | 18. Distribution Statement
Document is available to the public from the National Technical Information Service www.ntis.gov | |
| 19. Security Classif. (Of this report)
Unclassified | 20. Security Classif. (Of this page)
Unclassified | 21. No. of Pages
28 | 22. Price |

Form DOT F 1700.7 (8-72)   Reproduction of completed page authorized

Table of Contents

Executive Summary

This report analyzes the crash-reduction benefits of light-emitting diode (LED) stop lamps and LED center high-mounted stop lamps (CHMSL) using real-world crash data. Specifically, we sought an answer as to whether LED lamps were more beneficial than incandescent lamps at preventing rear-impact collisions.

Fifteen make-models were identified that had changed their stop lamps from all-incandescent to all-LED or vice-versa, a total of 17 switches counting make-models that changed more than once. However for these selected make-models, it was not possible to isolate the effect of LED stop lamps because each of these transitions coincided with a redesign of the rear-lighting configuration and/or a redesign of the entire vehicle such as a change in wheelbase or body style. Not a single make-model switched to LED stop lamps while leaving all other features of the vehicle unchanged. The Honda Accord coupe and sedan switched to LED in 2006 with a substantial reconfiguration of the rear-lighting system, although the rest of the vehicle stayed essentially the same. All the other make-models changed the rear-lighting configuration and the wheelbase when they switched to LED.

Using NHTSA's State Data System, involvement rates of LED-equipped versus incandescent-equipped vehicles as rear-ended vehicles in front-to-rear collisions with other vehicles were calculated for selected make-models that switched. This data was collected for each model's last two model years before and also the first two model years after the switch from all-incandescent to all-LED or vice-versa. Data from 12 States contained the variables needed to identify the relevant crashes and characteristics of the vehicles involved.

The main analysis method in this report compared the overall ratio of rear-impact crashes to a control group of frontal impacts before and after the switch to LED. A supplementary non-parametric analysis computed this ratio separately for each make-model and then compared the number of make-models that had lower rear-impact rates after the switch to the number of make-models that had higher rates.

The main analysis for the 2006 Honda Accord coupe and sedan alone found a statistically significant 7.3 percent reduction in rear impacts. When that analysis is extended to include the 15 other makes and models where the switch to LED was accompanied by a redesign of the entire vehicle, there was still a statistically significant 3.6 percent reduction in rear impacts. However, the non-parametric analysis does not show crash reduction for significantly more than 50 percent of the make-models. Instead, rear-impact rates actually increased with LED for the majority of these models: the main analysis showed a benefit only because the rates decreased with LED for the three models with the highest sales. Overall, these results do not support a conclusion about the effectiveness of LED stop lamps and LED CHMSL, but leave the question open for further study.

Introduction

There are alternatives to the incandescent stop lamps currently furnished on passenger vehicles. Previous research has studied alternatives such as LEDs, neon lamps, and a new type of fast-rising incandescent lamp. Research has shown that each of these other light sources has a faster rise time than current incandescent lamps – i.e., a shorter lag from the application of the brake until the light becomes visible to a human observer. Logically, if two lamps are equal in all other respects, the one with the faster rise time must have a safety benefit of alerting a following driver earlier that he or she must slow down and thereby reduce the number and severity of automobile crashes in situations requiring fast braking responses. But the benefit has not been quantified. It is also unknown if these new types of lamps have features (other than fast rise time) that make them more visible or less visible than incandescent lamps.

Over a million production vehicles have already been equipped with LED stop lamps, starting with the 2000 Cadillac Deville. This evaluation seeks to analyze and quantify the crash-reduction benefits of LED stop lamps using real-world crash data from NHTSA's State Data System by comparing the risk of rear-impact crashes of vehicles with LED stop lamps to the corresponding risk in vehicles of the same make and model with incandescent lamps. However, as we shall see, it was not possible to isolate the effect of LED stop lamps because, although numerous make-models switched from incandescent lamps in one year to LED lamps in the next, each of these transitions was accompanied by a major redesign of the vehicle, the rear-lighting configuration, or both.

Previous Research

Several laboratory experiments have been conducted measuring the time subjects took to respond to a brake light on a lead vehicle. Most of the experiments focused on the comparison between LED and incandescent technology. Viewed as a whole, these studies indicated a faster response time when the stop lamp was a LED unit. These studies established a theoretical basis for LED stop lamps (including the center high mounted stop lamp) being more beneficial than incandescent stop lamps.

The report provided by Fujita, Kouchi, Takahasji, and Kashiwabara in 1987[1] describes the technology and advantages of LED-based signal lamps. The main characteristic of LED-based signal lamps leading the campaign for being more beneficial than an incandescent source is the LED's faster rise-time. Rise-time is defined as the time interval from when the unit is energized to when it reaches a given level of output. The rise-time to 90% of full output is about 60 nanoseconds in the case of the LED, while it is about 140 milliseconds in the case of the incandescent source.

---

[1] Fujita, T., Kouchi, T., Takahasji, K., & Kashiwabara, H. (1987). *Development of LED High Mounted Stop Lamp.* (Paper No. 870061). Warrendale, PA: Society of Automotive Engineers.

This characteristic led to the project assessed by Olson[2] of UMTRI to determine whether the reaction-time advantage of LED units would be affected by various conditions that would be encountered in the real world. Under the conditions of the test that were most favorable for viewing signals, the LED units provided a response time advantage greater than expected, about 200 milliseconds. Under less favorable conditions, which was viewing at a distance and high-intensity illumination on the lamp surface, the response time of the LED units were less affected than those of the incandescent units and the response-time advantage increased to about 300 milliseconds.

A second UMTRI study,[3] which was similar to Olson's study, compared the braking response times for LED and incandescent stop lamps and also two alternatives, neon and fast incandescent lamps. The experiment tested all four different types of stop lamps at two levels of luminous intensity by placing a natural density filter in front and then not in front of the aperture. The neon, LED, and fast incandescent lamps all yielded shorter reaction times than did the standard incandescent lamp. Averaged over both levels of luminous intensity, the difference between the neon and LED lamps compared to the standard incandescent lamp averaged 166 milliseconds.

Goal of the Evaluation

The goal of this analysis is to determine the crash-reduction benefits of LED stop lamps using real-world crash data. All previous studies (none of which are based on crash data) came to the conclusion that the LED unit has a significant advantage over the incandescent unit in response time. This advantage suggests that the LED unit has greater conspicuity than an incandescent unit, which may be attributable to the LED's brief rise-time since the light will be visible sooner to the observer. With this information, it is important to test this theoretical basis that LED units will provide a safety benefit to help reduce the number and severity of automobile crashes in situations requiring fast braking responses.

Methods

To statistically test the hypothesis that LED stop lamps are more effective in crash reduction than the standard incandescent stop lamps, the first step is to identify crashes where the safety device is expected to have a benefit. Second, a group of crashes that serves as a control group – a measure of overall crash exposure in situations where the safety device should have no influence. Third is the concept of comparing crash involvement rates before and after the introduction of the safety device.

---

[2] Olson, P. L. (1987). *Evaluation of an LED High-Mounted Signal Lamp.* (UMTRI Technical Report No. UMTRI-87-13. Ann Arbor, MI: University of Michigan Transportation Research Institute. Available at http://deepblue.lib.umich.edu/bitstream/2027.42/34/2/74698.0001.001.pdf

[3] Sivak, M., Flannagan, M. J., Sato, T., Traube, E. C., & Aoki, M. (1993). *Reaction Times to Neon, LED, and Fast Incandescent Brake Lamps,* (UMTRI Technical Report 93-37). Ann Arbor, MI: University of Michigan Transportation Research Institute. Available at http://deepblue.lib.umich.edu/bitstream/2027.42/64045/1/84696.pdf

Crash Scenario

In all previous laboratory experiments, the setup generally placed the test lamps in front of the test subject. This layout concurs with the assumption that stop lamps are most influential in crashes when they are directly in front of the striking vehicle. With this reasoning, it is important to identify two-vehicle crashes where the struck vehicle has initial damage in the rear end at 5, 6 and 7 o'clock positions (as they are defined on FARS, or their equivalents on State files) while the striking vehicle has initial damage in the front end at 11, 12, and 1 o'clock positions.

This specific crash can also be taken a step further to where the stop lamps are the most conspicuous. These identified crashes can be limited to where a careful driver usually activates the stop lamps – slowing/stopping or stopped in traffic. There are also other situations where a careful driver may activate the stop lamps – turning left, turning right, making a U-turn, changing lanes, and merging into traffic. But due to the lower level of certainty that a careful driver may activate the stop lamps in these situations than when slowing/stopping or stopped in traffic, crashes involving these maneuvers for the struck vehicle were not considered.

Control Group

It is not appropriate to assume that all other crashes not in the test group provide a measure of crash exposure where the safety device should have no influence. It is important to realize that even though the stop lamps are not the most conspicuous, they may still be visible in a crash. An example of this is where the striking vehicle makes impact in the side rear (3, 4, 8, 9 o'clock positions) of the struck vehicle.

To ensure the device has no influence, it is appropriate to identify the control group as those vehicles in two vehicle crashes with initial damage in the front end at 11, 12, and 1 o'clock positions. It is not necessary to be specific with the damage area of the struck vehicle in this case.

Selection of Vehicle Models

The type of bulbs used in stop lamps and high-mount stop lamps were initially determined from two lists, one from NHTSA and another from North America Lighting, Inc. However, the two lists exhibited discrepancies among make-models. It became necessary to confirm the data by checking and cross-referencing with each available make-model's owner's manual. Images were also catalogued and verified across multiple model years.

Vehicles selected for the study were only those that had their stop lamps and center high-mounted stop lamp switched from all-incandescent in one model year to all-LED in the next year, or vice-versa. Vehicles with a mixture of incandescent and LED, e.g., incandescent stop lamps and LED CHMSL or vice-versa were removed from the list. Honda Accord coupe and sedan switched twice, from incandescent to LED to incandescent. For each make-model, two model years had to be available on each side of the switch, creating two-year blocks of all-incandescent or all-LED. In total, there had to be data for four model years of a make-model. These model years did not have to be consecutive model years since some make-models switched to a mixture before switching completely to one source of lighting. Also one all-incandescent make-model was not produced for a model year before being reintroduced for sale as an all-LED model.

Only 15 make-models were verified to have switched bulb types in stop lamps and CHMSL. Figure 1 shows the make-models and model years that are included in the study. A run of four model years (not necessarily consecutive), two with incandescent and two with LED (not necessarily in that order) will be called a "switch pair." Honda Accord coupe and sedan each contribute two switch pairs, namely 2004/5-2006/7 and 2006/7-2008/9. Each of the other models contributes one switch pair. Thus, there are a total of 17 switch pairs.

Figure 1: Make-Models and Model Years included in study

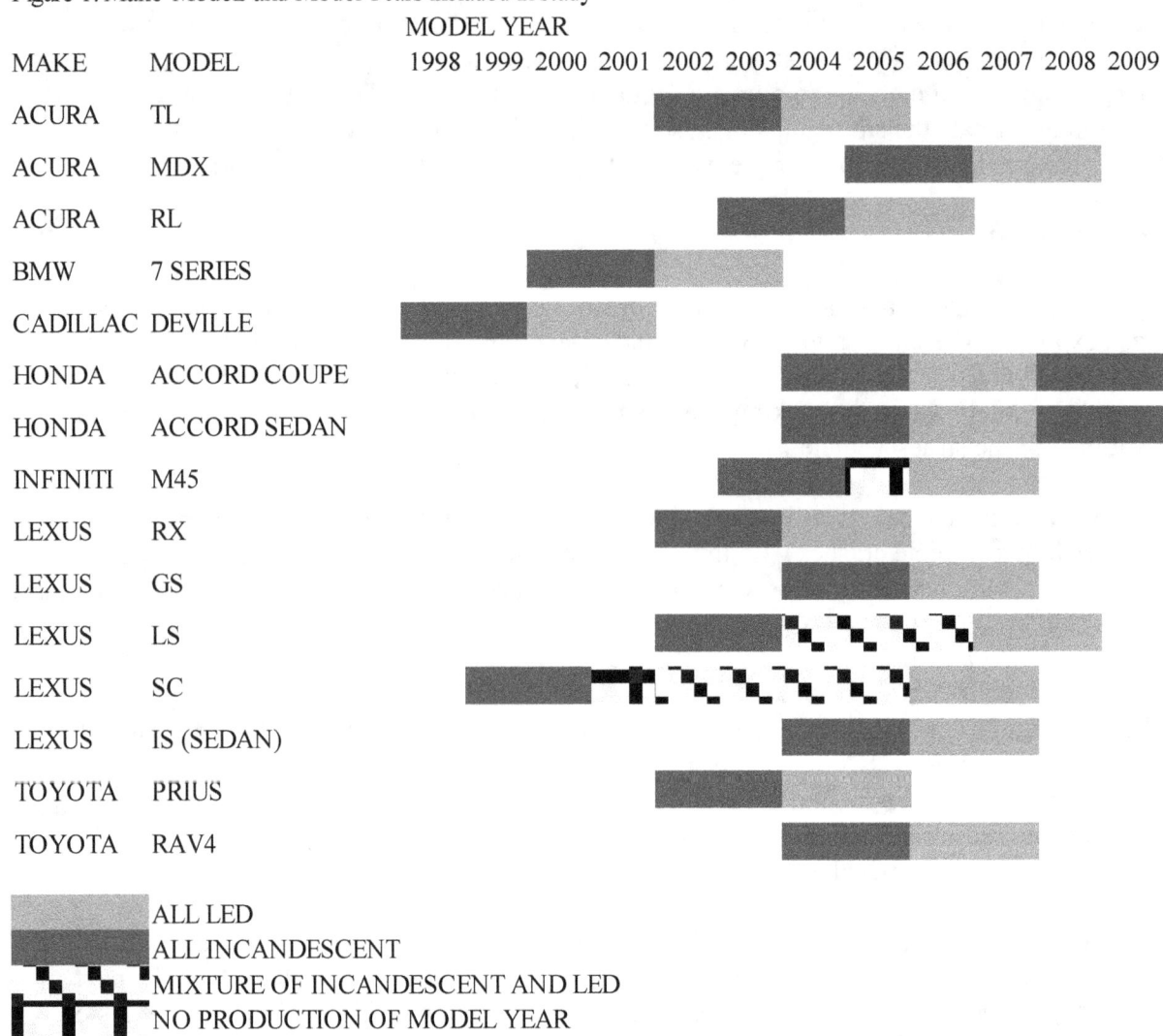

When selecting vehicles that completed a switch from all-incandescent to all-LED or vice-versa, the ideal was to search for "clean" switch pairs in which nothing changed except the shift from incandescent to LED or vice-versa. The concept behind a "clean" switch pair is that it controls

4

for driver and vehicle characteristics. If nothing else in the vehicle changes, any reduction in rear-impact crash rates may be attributed to the LED lamps.

Unfortunately, none of the 17 switch pairs was a "clean" switch pair. The switch from all-incandescent to all-LED usually coincided with a generation change (major redesign) in the vehicle, as evidenced by a change in the wheelbase or body style or a revision of the VIN. Out of the 15 make-models only the Honda Accord coupe and sedan remained in the same generation, and that only during the switch from incandescent to LED in MY 04-07. However, when the Honda Accord coupe and sedan switched back from all-LED to all-incandescent in MY 2008, it coincided with a generation change.

Figure 2 provides rear views of the Honda Accord coupe and sedan before and after the switch from incandescent to LED in 2006.

Figure 2: Honda Accord Coupe and Sedan, MY 2005 vs. MY 2006

Light Source	Incandescent		LED
Honda Accord Coupe MY 2005		MY 2006	
Wheelbase	105.1		105.1
Honda Accord Sedan MY 2005		MY 2006	
Wheelbase	107.9		107.9

Even though the Honda Accords stay in the same generation, it is evident that they are not "clean" switch pairs because the switch to all-LED was accompanied by a revision of the rear-lighting configuration. For the coupe, the CHMSL changes from being inside the car to the outside on the trunk. The sedan changes more extensively from having lights/reflectors on the trunk to only having the lamps on the side. And like the coupe, the CHMSL changes location from inside to outside for the sedan. All these changes might affect the conspicuity of the lamps and how drivers of the following vehicle react to them.

Staying with the Honda Accord, Figure 3 presents the switch from all-LED back to all-incandescent in MY 2008, which was accompanied with a generation change.

Figure 3: Honda Accord Coupe and Sedan, MY 2007 vs. MY 2008

Light Source	Incandescent		LED
Honda Accord Coupe MY 2008		MY 2007	
Wheelbase	107.9		105.1
Honda Accord Sedan MY 2008		MY 2007	
Wheelbase	110.2		107.9

This switch involves not only a change in wheelbase but also some reconfiguration of the stop lamps and CHMSL. For both the coupe and sedan, the CHMSL reverts in position from outside on the trunk to inside the car. The position of the CHMSL may affect the conspicuity of the lamps and how drivers of the following vehicle react to them.

The following images of the 13 remaining make-models in Figure 4 provide visual evidence of how the switch from all-incandescent to all-LED was accompanied by a major redesign of the vehicle, the rear-lighting configuration, or both, which may have changed the conspicuity of the lamps and how drivers of the following vehicle react to them.

Figure 4: Visuals of the Switch for the Remaining 13 Make-Models

Light Source	Incandescent		LED
Acura TL MY 2003		MY 2004	
Wheelbase	108.1		107.9

Light Source	Incandescent		LED
Acura RL MY 2004		MY 2005	
Wheelbase	114.6		110.2
Acura MDX MY 2006		MY 2007	
Wheelbase	106.3		108.3
BMW 7 Series MY 2001		MY 2002	
Wheelbase	115.4 / 120.9		117.7 / 123.2
Cadillac Deville MY 1999		MY 2000	
Wheelbase	113.8		115.3 / 115.6
Infiniti M45 MY 2004		MY 2006	
Wheelbase	113.8		115.3 / 115.6

Light Source	Incandescent		LED
Lexus RX MY 2003		MY 2004	
Wheelbase	103.0		106.9
Lexus GS MY 2005		MY 2006	
Wheelbase	110.2		112.2
Lexus LS MY 2003		MY 2007	
Wheelbase	115.2		116.9
Lexus IS (Sedan) MY 2005		MY 2006	
Wheelbase	105.1		107.5
Lexus SC MY 2000		MY 2006	
Wheelbase	105.9		103.1

Light Source	Incandescent		LED
Toyota Prius MY 2003		MY 2004	
Wheelbase	100.4		106.3
Toyota RAV4 MY 2005		MY 2006	
Wheelbase	98		104.7

Out of the 13 make-models, there were noticeable changes that occurred during the switch involving the design of the vehicle, the rear-lighting configuration, or both. All make-models during the switch were accompanied with a change in wheelbase. The Toyota Prius and the Lexus SC provide examples of how the switch accompanies a major redesign of the vehicle. The Toyota Prius starts out as a sedan, but after the switch is now a hatchback. The Lexus SC switches from a coupe to a convertible during the switch. The Acura MDX is a good example of switches involving redesigns in rear-lighting configuration. The switch to LED allows for the rear tail lamps to extend over onto the trunk, while decreasing its vertical length. Another example is the Lexus RX. The LED version has a clear outer lens and reflective surfaces on the housing that may make viewing the rear lighting more difficult in the daytime. Finally, the Cadillac Deville provides an example of how the switch from incandescent to LED involved a major redesign to both the vehicle and the rear-lighting configuration. The Cadillac Deville provides a new design to the body style of the vehicle, while also widening the rear tail lamps and changing the location of CHMSL from inside the vehicle to the outside on the trunk. Even though most of these changes in body style and/or rear-lighting configuration are not drastic, they could still affect to some extent the ability of drivers of the following vehicle to react to the lamps or they could affect what sort of people buy the vehicles and what sort of crashes they get into. It is also worthwhile to note that most of these lamp switches not only included a change from incandescent to LED, but also a change from a 1-lighted section lamp to a 3-lighted section lamp. FMVSS No. 108 requires/allows ~40% more light output from the latter. These changes that occurred during the switch from all-incandescent to all-LED shows how hard it is to isolate the sole effect of LED stop lamps in the absence of "clean" switch pairs.

By contrast, NHTSA's 2009 evaluation of amber rear turn signals had a database of 33 switch pairs, of which 23 did not involve a generational change in the make-model and 11 were "clean" switch pairs in which the rear-lighting configuration appeared to stay exactly the same except for the switch from red to amber lenses in the turn signals.[4]

The 2×2 Contingency Table

The analysis objective is to compare the involvement rates of LED-equipped vs. incandescent-equipped vehicles as rear-impacted vehicles in front-to-rear collisions with other vehicles. The control group is the same make and model as a frontally impacting vehicle in collisions with other vehicles. Involvements in crashes for a group of make-models that switched from incandescent to LED stop lamps are tabulated by vehicle type (Incandescent; LED) and crash type (rear impact; front impact). The 2×2 contingency table is:

Number of Crash Involvments

Type of Car	Rear Impacts	Front Impacts (Control Group)
Incandescent	N_{11}	N_{12}
LED	N_{21}	N_{22}

The number of control-group involvements is a surrogate for the "exposure" of a group of vehicles. The LED-equipped vehicles have N_{22} / N_{12} times as much exposure as the incandescent-equipped vehicles. Based on this exposure ratio, the expected number of rear impacts in the LED-equipped vehicles is $(N_{22} / N_{12}) \times N_{11}$. In fact, there are only N_{21} rear impacts in the LED-equipped vehicles. Using this information provides a reduction of

$$1 - [(N_{21}/N_{11})/(N_{22}/N_{12})]$$

in the rear-impact involvement rate, which yields a point estimate of the effectiveness or simply "LED effectiveness." The results of the "LED effectiveness" can only be interpreted in three ways. A value of 0.00 means that the two vehicle types were equally likely to be involved in a collision. A value greater than 0.00 means that the LED-equipped vehicle was less likely to experience the event. A value less than 0.00 means that the LED-equipped vehicle was more likely to experience the event. With the values in the contingency table representing counts and not proportions or averages, the "LED effectiveness" can be multiplied by 100 to provide a percent effectiveness.

The chi-square (χ^2) test was used to assess the statistical significance of the relationship in the contingency table since the table's cells represent counts and the values of the expected frequencies were large (5 or greater). A χ^2 of 3.84 or larger leads to the rejection of the null hypothesis and the conclusion that the involvement rate varies by vehicle type.

---

[4] Allen, K. (2009). *The Effectiveness of Amber Rear Turn Signals for Reducing Rear Impacts.* (Report No. DOT HS 811 115). Washington, DC: National Highway Traffic Safety Administration. Available at www-nrd nhtsa.dot.gov/Pubs/811115.PDF

Non-Parametric Analysis

A non-parametric analysis might or might not provide additional evidence that the results provided from the contingency table analysis captured the sole effects of the switch from incandescent to LED or vice-versa. The non-parametric analysis uses the same crash data, but computes effectiveness separately for each switch pair and then simply compares the number of switch pairs that had lower rear-impact crash rates after the switch to LED to the number of switch pairs that had higher rates. In other words, did significantly more than half the switch pairs improve (have lower rates with LED)? Advantages of this method are: (1) It "controls" for make and model; it is not influenced by some models having mostly "all-incandescent" cases and others have mostly "all-LED" cases; and (2) overall findings are not overly influenced by one or two high-sales make-models with anomalous results: all models have equal weight. The disadvantage from this analysis, though, is that it is less likely to produce significant results, from the same number of cases, than the principal method; thus, a non-significant finding is not necessarily a negative result, just a caution flag. But if the non-parametric analysis actually shows crash rates increasing with LED for the majority of switch pairs, it is a stronger caution flag. NHTSA's evaluation of amber rear turn signals, for example, used a non-parametric analysis to corroborate the basic risk-ratio analysis.

Data Sources

The analyses require large samples of crash data that specify VINs, impact locations, and vehicle maneuvers immediately prior to the collision. NHTSA's State Data System is the data source for this project. State files can furnish an adequate number of cases for statistical analyses. Currently, the agency receives data from 33 States and maintains them for analysis.

The critical parameters that must be defined in each State file – VIN, model year, initial impact location, vehicle maneuver immediately prior the collision – will now be discussed, in that order.

State Data System is compiled from police-reported crashes in a State. To ensure the vehicles selected were actually the switch vehicles, it was necessary that the make-models be identifiable from the VIN, based on decode programs developed by NHTSA staff for use in evaluations of Federal Motor Vehicle Safety Standards and other vehicle safety analyses. Using only States that record VIN data on the file they supply to NHTSA removes 9 States from the study: Arkansas, California, Colorado, Connecticut, Iowa, Minnesota, Montana, South Carolina, and Texas. Going one step further, another 4 States are removed from the analysis: Delaware, because VIN data is only collected for calendar years (CY) 2007-2008; Indiana, because VIN data that is collected from CY 03-08 is only for commercial vehicles; Ohio, because VIN data is not collected past CY 1999; Virginia, because VIN data is only collected for CY 2005-2006. This leaves 20 State files as candidates to be included in the analysis.

A consistency check on the data is that the VIN's model year code must match the model year reported by the police at the crash. This criterion removes Washington State since model year data is not collected past CY 1996. Illinois provides VIN data for CY 1989-2009, but for CY 2004-2006 there is no model year data to provide validity for the VINs. Since only two CYs are missing for the MY 1998-2009, it was accepted to use the valid VINs and derive the model year information from the VINs for CY 2004-2006 in Illinois.

Every State has its own unique way of coding a vehicle's impact location. "Rear impact" is not defined exactly the same way in each State, but at least we can make the definitions as similar as possible. But the State file must have some variable for a vehicle's impact location. Since the study is specifically looking at two vehicle crashes, there should be an impact location code specific to each vehicle involved in the collision. With this requirement, 3 more State files are excluded from the study: New York, New Mexico, and Wisconsin. It is also desirable to make the "rear impact" category strictly related to the initial point of impact to the vehicle and not necessarily the location of the most damage to the vehicle. We defined the "rear impact" category to be as exclusive as possible – any crashes where the impact encompasses at least part of the back portion of the side of the car are not to be included among the rear impacts, because it is important to analyze crashes where there is a very high certainty that the stop lamps and CHMSL were clearly visible to the striking vehicle. Thus, it is necessary to identify the initial impact location of the striking vehicle to be only on the front end and no front portion of the side of the car. This desired collision type is purely of a collision occurring on a straight line where the striking vehicles makes impact into the back of the struck vehicle. For each State, using the clock as a point of reference, rear impact tried to include damage to the struck vehicle strictly at the 5, 6, and 7 o'clock positions with damage to the striking strictly at the 11, 12, and 1 o'clock positions.

A second analysis focuses solely on front to rear impacts in which the struck vehicle is performing the vehicle maneuver immediately prior to the collision of slowing/stopping or stopped in traffic. This maneuver identifies crashes in which the stop lamps and CHMSL are assumed to be active. Similar to vehicle's impact location, every State has its own unique way of coding a vehicle's prior maneuver to the collision. Missouri provides three action codes to describe the vehicle's maneuver. The first vehicle's maneuver code is used since it provides the most responses out of the three action codes. In each State, vehicle maneuver included actions that involved slowing, stopping, or being stopped, but excluded making right turns, left turns or U-turns, entering/leaving parked position, merging or changing lanes.

State-by-State, the following data count for the selected 15 switch vehicles when being struck in a front-to-rear collision and while slowing/stopping/stopped and being struck in front-to-rear:

State	Data Count for Front-to-Rear	Data Count for Front-to-Rear While Struck Vehicle is Slowing/Stopping/Stopped
Alabama	9,066	7,696
Florida	15,614	13,921
Georgia	24,780	19,187
Illinois	31,542	25,163
Kansas	1	1
Kentucky	7,308	6,158
Maryland	4,931	4,463
Michigan	9,225	7,972
Missouri	5,697	4,302
Nebraska	1,542	1,230
New Jersey	41,588	34,674
North Carolina	6,809	5,954
North Dakota	162	29
Pennsylvania	8,231	7,148
Utah	378	316
Wyoming	250	216

From the data count, it is obvious that Kansas, North Dakota, Utah, and Wyoming provide little data and should be excluded from the study.

For this analysis, data is only collected for all model years less than or equal to the most recent calendar year collected by each State. On top of that, contingency tables are only created if data is available for all model years before and after the switch. For example, if a vehicle switched in MY 2008, data would be collected for MY 2006-2009. But if no data is available for MY 2009 then no two-MY switch contingency table can be created. The remaining 12 State files were used in the analysis for the calendar years available to NHTSA in which VINS were reported:

State	Calendar Years	Model Years to Exclude
Alabama	1995-2008	2009-2010
Florida	1993-2009	2010
Georgia	1989-1990 & 1998-2008	2009-2010
Illinois	1989-2009	2010
Kentucky	1997-2009	2010
Maryland	1989-2008	2009-2010
Michigan	1989-1991 & 2004-2009	2010
Missouri	1989-2008	2009-2010
Nebraska	1999-2008	2009-2010
New Jersey	2001-2010	none
North Carolina	1992-2006	2007-2010
Pennsylvania	1989-2001 & 2003-2008	2009-2010

Results

Analysis of 2004-2007 Honda Accord

As stated in the *Selection of Vehicle Models* section, the Honda Accord's switch from incandescent lamps in 2004-2005 to LED in 2006-2007 has been the only switch so far that did not coincide with a redesign of the vehicle and a change in wheelbase. However, it did coincide with substantial changes in the rear-lighting configuration. Table 1 provides the 2×2 contingency table as illustrated before for the coupe and sedan individually and together with the test group strictly rear-to-front collisions. Table 2 provides similar information, but limits the test group to rear-to-front collisions where the struck vehicle is slowing/stopping or is stopped in traffic.

Table 1: Honda Accord 2×2 Table for Rear-to-Front Collisions

Coupe		Rear Impacts	Front Impacts	Odds
	Incandescent	2,005	3,080	65.10%
	LED	707	1,312	53.89%
				17.2% Effectiveness
	$\chi^2 = 11.92$		**P-Value < 0.01**	

Sedan		Rear Impacts	Front Impacts	Odds
	Incandescent	10,229	14,063	72.74%
	LED	5,825	8,566	68.00%
				6.5% Effectiveness
	$\chi^2 = 9.91$		**P-Value < 0.01**	

Both		Rear Impacts	Front Impacts	Odds
	Incandescent	12,234	17,143	71.36%
	LED	6,532	9,878	66.13%
				7.3% Effectiveness
	$\chi^2 = 14.73$		**P-Value < 0.01**	

Table 2: Honda Accord 2×2 Table for Rear-to-front Collisions With Struck Vehicle Is Slowing/Stopping or Stopped in Traffic

Coupe		Rear Impacts	Front Impacts	Odds
	Incandescent	1,163	3,080	37.76%
	LED	400	1,312	30.49%
				19.3% Effectiveness
	$\chi^2 = 10.31$		**P-Value < 0.01**	

Sedan		Rear Impacts	Front Impacts	Odds
	Incandescent	6,165	14,063	45.63%
	LED	3,382	8,566	30.26%
				9.9% Effectiveness
	$\chi^2 = 16.97$		**P-Value < 0.01**	

Both	Rear Impacts	Front Impacts	Odds
Incandescent	7,328	17,143	42.75%
LED	3,782	9,878	38.29%
			10.4% Effectiveness

$$\chi^2 = 21.67 \qquad \text{P-Value} < 0.01$$

The overall result for all front to rear collisions is a 7.3 percent reduction of rear impacts, which is statistically significant with $\chi^2 = 14.73$ and the p-value < 0.01. Rear-to-front with struck vehicle slowing/stopping/stopped resulted in a 10.4 percent crash reduction that was statistically significant with $\chi^2 = 21.67$ and the p-value < 0.01. It is a good sign to see that the crash reduction improved once limiting all rear-to-front crashes to when the struck vehicle is more likely to be braking. However, there are also caution flags raised due to the high reduction percentages – perhaps "too good to be true," when compared to the introduction of CHMSL in 1986, which reduced rear-impact risk by just 4.3 percent, even though it may have been a more substantial change in rear-lighting than the shift from incandescent to LED.[5]

A possible factor that may explain the results could be vehicle-age effect, due to LED-equipped vehicles being newer than the incandescent-equipped vehicles. The ratio of rear-to-frontal impacts may change as vehicles age and are driven by different people. It is essential to ensure that the results provided above are not biased due to vehicle age. Table 3 checks for possible vehicle-age effects by performing a similar analysis as Table 1, but now comparing all-incandescent two model years before or three model years before the shift to LED to all-incandescent one model year before the shift to LED.

Table 3: Honda Accord – Possible Vehicle Age Effect for Rear-to-front Collisions
(Comparison of all-incandescent to all-incandescent)

MY	Rear Impacts	Front Impacts	Odds
2004	7,385	10,381	71.14%
2005	5,577	7,801	71.49%
			-0.5% Effectiveness

$$\chi^2 = 0.04 \qquad \text{P-Value} = 0.83$$

MY	Rear Impacts	Front Impacts	Odds
2003	9,613	13,953	68.90%
2005	5,577	7,801	71.49%
			-3.8% Effectiveness

$$\chi^2 = 2.83 \qquad \text{P-Value} = 0.09$$

The results all show non-significant year-to-year differences in the ratio of rear-to-front impact for the model years before the switch to LED. Thus, the large reduction in the year they switched to LED is not likely due to a vehicle-age effect. While it sets aside the issue of vehicle-

---

[5] Kahane, C. J., & Hertz, E. (1998). *The Long-Term Effectiveness of Center High Mounted Stop Lamps in Passenger Cars and Light Trucks*. (Report No. DOT HS 808 696). Washington, DC: National Highway Traffic Safety Administration. Available at www-nrd.nhtsa.dot.gov/pubs/808696.pdf

age bias, Table 3 does not address other issues, such as the effect of the change in the rear-lighting configuration, or even the possibility of an unexplained market shift for Honda Accord in 2006 that resulted in a different group of drivers with different crash patterns.

Extended Analysis Including Vehicles With Generation Change

Because it is undesirable to rely on an analysis of just two switch pairs, it is necessary to bring other switch pairs into the analysis even though the switch coincided with a generation change. All the other 15 switch pairs are included in this analysis along with the preceding results in Tables 1 and 2 for the Honda Accord Coupe and the Honda Accord Sedan, totaling to 17 switch pairs. Table 4 and 5 provide the results of the contingency tables for the other 15 individual switch pairs. Table 6 and 7 provide the overall, combined results for all vehicles in the 17 switch pairs, based on the crash data from all 12 States.

Table 4: Generation Change 2×2 Table for Rear-to-front Collisions

Acura TL	Rear Impacts	Front Impacts	Odds
Incandescent	3,382	5,752	58.80%
LED	2,338	3,102	75.37%
			-28.2% Effectiveness
	$\chi^2 = 50.65$	**P-Value < 0.01**	

Acura MDX	Rear Impacts	Front Impacts	Odds
Incandescent	1,559	1,622	96.12%
LED	698	629	110.97%
			-15.5% Effectiveness
	$\chi^2 = 4.83$	**P-Value = 0.03**	

Acura RL	Rear Impacts	Front Impacts	Odds
Incandescent	179	320	55.94%
LED	283	524	54.01%
			3.5% Effectiveness
	$\chi^2 = 0.09$	P-Value = 0.77	

BMW 7 Series	Rear Impacts	Front Impacts	Odds
Incandescent	828	1,023	80.94%
LED	610	808	75.50%
			6.7% Effectiveness
	$\chi^2 = 0.96$	P-Value = 0.33	

Cadillac Deville	Rear Impacts	Front Impacts	Odds
Incandescent	4,893	7,799	62.74%
LED	2,902	6,215	46.69%
			25.6% Effectiveness
	$\chi^2 = 104.36$	**P-Value < 0.01**	

16

Honda Accord Coupe[6]	Rear Impacts	Front Impacts	Odds
Incandescent	340	458	74.24%
LED	466	899	51.84%
			30.2% Effectiveness
	$\chi^2 = 15.44$	**P-Value < 0.01**	

Honda Accord Sedan[7]	Rear Impacts	Front Impacts	Odds
Incandescent	1,792	2,064	86.82%
LED	3,752	5,855	64.08%
			26.2% Effectiveness
	$\chi^2 = 62.51$	**P-Value < 0.01**	

Infiniti M45	Rear Impacts	Front Impacts	Odds
Incandescent	153	194	78.87%
LED	533	634	84.07%
			-6.6% Effectiveness
	$\chi^2 = 0.27$	P-Value = 0.60	

Lexus RX	Rear Impacts	Front Impacts	Odds
Incandescent	2,175	2,608	83.40%
LED	3,225	3,860	83.55%
			-0.2% Effectiveness
	$\chi^2 < 0.01$	P-Value = 0.96	

Lexus GS	Rear Impacts	Front Impacts	Odds
Incandescent	110	141	78.01%
LED	576	610	94.43%
			-21.0% Effectiveness
	$\chi^2 = 1.87$	P-Value = 0.17	

Lexus LS	Rear Impacts	Front Impacts	Odds
Incandescent	766	888	86.26%
LED	195	295	66.10%
			23.4% Effectiveness
	$\chi^2 = 6.49$	**P-Value = 0.01**	

Lexus SC	Rear Impacts	Front Impacts	Odds
Incandescent	38	118	32.20%
LED	40	68	58.82%
			-82.7% Effectiveness
	$\chi^2 = 4.93$	**P-Value = 0.03**	

---

[6] The switch from LED lamps in 2006-2007 to incandescent in 2008-2009.
[7] Same as footnote 6.

Lexus IS (Sedan)	Rear Impacts	Front Impacts	Odds	
Incandescent	184	282	65.25%	
LED	872	1,033	84.41%	
			-29.4% Effectiveness	
	$\chi^2 = 6.00$	**P-Value = 0.01**		

Toyota Prius	Rear Impacts	Front Impacts	Odds	
Incandescent	331	577	57.37%	
LED	1,143	1,759	64.98%	
			-13.3% Effectiveness	
	$\chi^2 = 2.51$	P-Value = 0.11		

Toyota RAV4	Rear Impacts	Front Impacts	Odds	
Incandescent	2,858	2,800	102.07%	
LED	3,129	2,808	111.43%	
			-9.2% Effectiveness	
	$\chi^2 = 5.57$	**P-Value = 0.02**		

Table 5: Generation Change 2×2 Table for Rear-to-front Collisions With Struck Vehicle Is Slowing/Stopping or Stopped in Traffic

Acura TL	Rear Impacts	Front Impacts	Odds	
Incandescent	2,032	5,752	35.33%	
LED	1,429	3,102	46.07%	
			-30.4% Effectiveness	
	$\chi^2 = 41.84$	**P-Value < 0.01**		

Acura MDX	Rear Impacts	Front Impacts	Odds	
Incandescent	985	1,622	60.73%	
LED	434	629	69.00%	
			-13.6% Effectiveness	
	$\chi^2 = 2.95$	P-Value = 0.09		

Acura RL	Rear Impacts	Front Impacts	Odds	
Incandescent	99	320	30.94%	
LED	169	524	32.25%	
			-4.3% Effectiveness	
	$\chi^2 = 0.08$	P-Value = 0.77		

BMW 7 Series	Rear Impacts	Front Impacts	Odds	
Incandescent	502	1,023	49.07%	
LED	342	808	42.33%	
			13.7% Effectiveness	
	$\chi^2 = 3.07$	P-Value = 0.08		

Cadillac Deville	Rear Impacts	Front Impacts	Odds
Incandescent	2,644	7,799	33.90%
LED	1,513	6,215	24.34%
			28.2% Effectiveness
	$\chi^2 = 82.94$	**P-Value < 0.01**	

Honda Accord Coupe[8]	Rear Impacts	Front Impacts	Odds
Incandescent	209	458	45.63%
LED	272	899	30.26%
			33.7% Effectiveness
	$\chi^2 = 14.45$	**P-Value < 0.01**	

Honda Accord Sedan[9]	Rear Impacts	Front Impacts	Odds
Incandescent	1,155	2,064	55.96%
LED	2,252	5,855	38.46%
			31.3% Effectiveness
	$\chi^2 = 71.92$	**P-Value < 0.01**	

Infiniti M45	Rear Impacts	Front Impacts	Odds
Incandescent	86	194	44.33%
LED	297	634	46.85%
			-5.7% Effectiveness
	$\chi^2 = 0.14$	P-Value = 0.71	

Lexus RX	Rear Impacts	Front Impacts	Odds
Incandescent	1,350	2,608	51.76%
LED	1,941	3,860	50.28%
			2.9% Effectiveness
	$\chi^2 = 0.44$	P-Value = 0.51	

Lexus GS	Rear Impacts	Front Impacts	Odds
Incandescent	59	141	41.84%
LED	350	610	57.38%
			-37.1% Effectiveness
	$\chi^2 = 3.51$	P-Value = 0.06	

Lexus LS	Rear Impacts	Front Impacts	Odds
Incandescent	433	888	48.76%
LED	122	295	41.36%
			15.2% Effectiveness
	$\chi^2 = 1.81$	P-Value = 0.18	

---

[8] Same as footnote 6.
[9] Same as footnote 6.

Lexus SC	Rear Impacts	Front Impacts	Odds
Incandescent	24	118	20.34%
LED	29	68	42.65%
			-109.7% Effectiveness
	$\chi^2 = 5.64$	**P-Value = 0.02**	

Lexus IS (Sedan)	Rear Impacts	Front Impacts	Odds
Incandescent	126	282	44.68%
LED	528	1,033	51.11%
			-14.4% Effectiveness
	$\chi^2 = 1.26$	P-Value = 0.26	

Toyota Prius	Rear Impacts	Front Impacts	Odds
Incandescent	175	577	30.33%
LED	661	1,759	37.58%
			-23.9% Effectiveness
	$\chi^2 = 4.83$	**P-Value = 0.03**	

Toyota RAV4	Rear Impacts	Front Impacts	Odds
Incandescent	1,711	2,800	61.11%
LED	1,952	2,808	69.52%
			-13.8% Effectiveness
	$\chi^2 = 9.19$	**P-Value < 0.01**	

Table 6: All Vehicles 2×2 Table for Rear-to-Front Collisions

All Vehicles	Rear Impacts	Front Impacts	Odds
Incandescent	31,822	43,789	72.67%
LED	27,294	38,977	70.03%
			3.6% Effectiveness
	$\chi^2 = 11.80$	**P-Value < 0.01**	

Table 7: All Vehicles 2×2 Table for Rear-to-Front Collisions With Struck Vehicle Is Slowing/Stopping or Stopped in Traffic

All Vehicles	Rear Impacts	Front Impacts	Odds
Incandescent	18,918	43,789	43.20%
LED	16,073	38,977	41.24%
			4.6% Effectiveness
	$\chi^2 = 13.26$	**P-Value < 0.01**	

The results for all 17 switch pairs provided a statistically significant reduction of 3.6 percent crash reduction in all rear-to-front collisions. Rear-to-front with struck vehicle slowing/stopping/stopped resulted in a 4.6 percent crash reduction that was statistically significant with $\chi^2 = 13.26$ and the p-value < 0.01.

Table 8 and 9 present the results of two non-parametric analyses of all 17 switch pairs, one for all rear-to-front collisions and one for rear impacts where the vehicle was slowing, stopping, or stopped in traffic. In both analyses, only 8 switch pairs had lower rear-impact crash rates with LED, while 9 switch pairs become worse with LED. A ratio of 9 to 8 is still within the binomial test's acceptance range for the null hypothesis of a 50/50 split; thus, at least it does not imply that LED increased risk, but in any case it certainly does not support the conclusion that LED is effective, as risk increased for the majority of the make-models. This shows that even though the analyses of crash involvement rates for all vehicles resulted in positive effectiveness (because LED was effective on the highest-sales make-models such as Honda Accord), that result may be questioned because the majority of switch pairs did not improve with LED.

Table 8: Non-Parametric Analysis Table for Rear-to-Front Collisions

IMPROVED	8
WORSEN	9
TOTAL	17

Table 9: Non-Parametric Analysis Table for Rear-to-Front Collisions With Struck Vehicle Slowing/Stopping or Stopped in Traffic

IMPROVED	8
WORSEN	9
TOTAL	17

By contrast, the non-parametric analysis in NHTSA's evaluation of amber rear turn signals showed an improvement with amber turn signals in 24 of its 33 switch pairs, while only 9 did worse with amber. This was an improvement in significantly more than half the switch pairs, based on the binomial test, and it strongly corroborated the findings from the risk-ratio analysis that amber turn signals were effective.

Discussion

The contingency-table analyses of crash reduction yielded statistically significant positive effectiveness for LED-equipped vehicles in rear impacts, both for the 2006-2007 Honda Accord alone and for an extended analysis that included 15 other makes and models where the switch to LED was accompanied by a vehicle redesign. But the results of this analysis are not at all corroborated by the nonparametric analysis, which actually showed risk increasing with LED for a majority of the individual make-models. We are unable to draw any meaningful conclusion because our database did not include a single make-model that switched from incandescent to LED or vice-versa without other simultaneous major changes. The study could not analytically isolate for the change in light source, but may have captured other factors that occurred to the make/models during the switch from incandescent to LED. These other factors include the potential effects of major changes in the rear-lighting configuration (such as moving the CHMSL to a more prominent location) as well as the consequences of major vehicle redesigns, which can change both how the vehicle performs and what sort of people buy it. As of now, the real-world crash data does not demonstrate that LED stop lamps and LED CHMSL are more beneficial than incandescent lamps, but also fail to rule out such a possibility. It would be a good idea to revisit the issue sometime in the future when there are a number of "clean" switch pairs or, at least, a larger number of reasonably high-sales switch pairs than there have been to date – i.e., when there is a database more comparable to the 34 switch pairs, including 11 "clean" switch pairs available for NHTSA's analysis of amber rear turn signals.

DOT HS 811 712
February 2013

U.S. Department
of Transportation

**National Highway
Traffic Safety
Administration**

9335-020113-v2